sea fever

NATURISTS ON STUDLAND BEACH, DORSET, 2005

sea fever

stuart franklin

THE
BARDWELL
PRESS

Published by:

The Bardwell Press
6 Bardwell Road
Oxford OX2 6SW
www.bardwell-press.co.uk

www.magnumphotos.com

British Library Cataloguing in Publication Data
A catalogue record for this book is available from the British Library

ISBN 0-9548683-4-X

COVER PHOTOGRAPH: CENTENARY BALL, BRITANNIA ROYAL NAVAL COLLEGE, DARTMOUTH, 2005

MAGNUM PHOTOS

THE NATIONAL TRUST

Designed and Typeset by The Bardwell Press, Oxford.
Printed in Italy by EBS Verona

contents

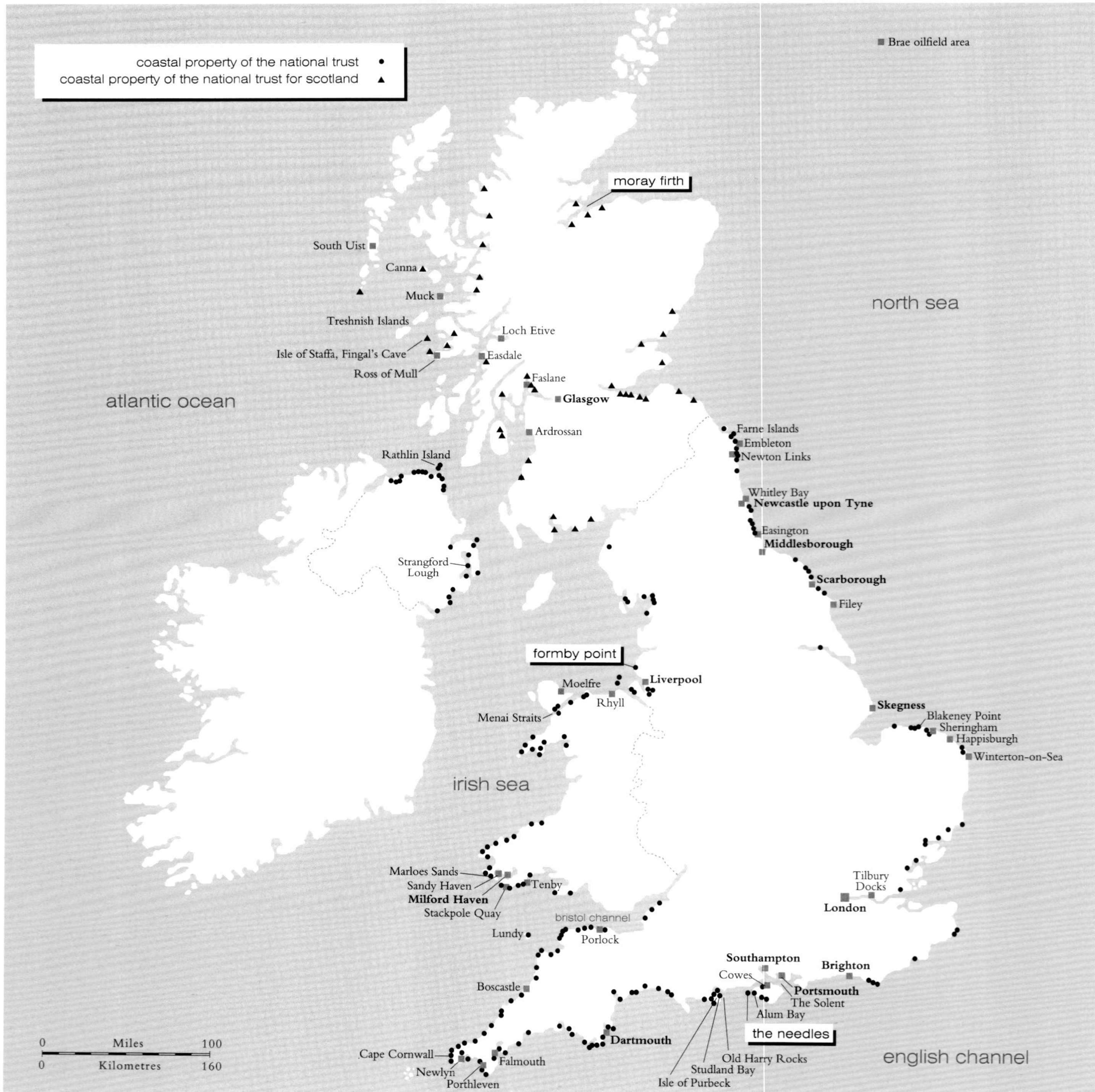
coastal property of the national trust
coastal property of the national trust for scotland
Brae oilfield area
moray firth
South Uist
Canna
Muck
Treshnish Islands
Loch Etive
Isle of Staffa, Fingal's Cave
Easdale
Ross of Mull
Faslane
Glasgow
north sea
atlantic ocean
Ardrossan
Farne Islands
Embleton
Newton Links
Rathlin Island
Whitley Bay
Newcastle upon Tyne
Easington
Middlesborough
Strangford Lough
Scarborough
Filey
formby point
Liverpool
Moelfre
Rhyll
Menai Straits
Skegness
Blakeney Point
Sheringham
Happisburgh
Winterton-on-Sea
irish sea
Tilbury Docks
London
Marloes Sands
Sandy Haven
Milford Haven
Stackpole Quay
Tenby
bristol channel
Lundy
Porlock
Southampton
Brighton
Cowes
Portsmouth
The Solent
Alum Bay
Boscastle
the needles
Dartmouth
Cape Cornwall
Falmouth
Newlyn
Porthleven
Old Harry Rocks
Studland Bay
Isle of Purbeck
english channel
0 Miles 100
0 Kilometres 160

foreword

This book is all about celebrating the beauty and diversity of our coast, expressed through stunning images—I defy anyone not to be inspired by the pictures in this collection. They are why we were so keen to promote *The Coast Exposed* exhibition: to celebrate the beauty, diversity and wildlife of the British coast, and provide an intimate record of the relationship between the coast, those who live and work there and all those who enjoy it.

Sea Fever is published in a year of anniversaries linked to our maritime heritage and coast, not least Nelson's victory at Trafalgar. The National Trust is proud to play a part in SeaBritain and Sea Northern Ireland 2005. The Trust has maritime connections with sea-faring people, past and present, through our houses, coast and countryside, chattels, paintings and furniture which continue to enthral and fascinate our visitors—from the entire collection of the Gibbs family at Tyntesfield, whose wealth came from the guano trade with Peru and who once owned *SS Great Britain*, to our own passenger vessel, the *Gondola*, that plies Lake Coniston in the shadow of the Lakeland fells.

One anniversary in 2005 is particularly special to the National Trust—that of our own Neptune Coastline Campaign. Forty years ago much of our coastline was disappearing under commercial development and its future was vulnerable. Something had to be done, and we founded Neptune in response. With tremendous support from the public we have been able to buy, manage and provide access to over 700 miles of stunning coastline. Launching the campaign in 1965 HRH The Duke of Edinburgh said it was "a lifebelt just within reach"; its founders never dreamt it would be so successful.

But, as ever, the challenges of the future are if anything greater. Across the Trust's coastal holdings, the pressures loom large; in particular climate change and sea level rise. We are often the first to feel the impacts of change and we are committed to using our own direct experience and expertise to show how, working with nature, not against it, we can help promote a more sustainable future.

I am sure this beautiful book will stand as a valuable record of our evolving coast, and the dynamic natural and cultural relationships it chronicles.

Fiona Reynolds, Director-General of the National Trust, August 2005

DOLPHINS ON THE BOW, BRISTOL CHANNEL, 2005

introduction | sea fever

I must down to the seas again, for the call of the running tide
Is a wild call and a clear call that may not be denied.

FROM *SEA-FEVER* BY JOHN MASEFIELD

for most of my professional life I have worked and travelled overseas, mainly—for the past 15 years—on assignment for *National Geographic Magazine*. I regarded Britain as cold, grey and dull compared with the exotic charm of the Inca Trail to Macchu Pichu, Saudi Arabia's Empty Quarter, or the rice terraces of Bali.

In June 2004 I was commissioned by the National Trust to contribute towards an exhibition on the coast as part of SeaBritain 2005, a festival loosely centred on the 200th anniversary of Nelson's victory at the Battle of Trafalgar. During five days spent on the Farne Islands and along the Northumberland coast I became intrigued by several aspects of the British coast: great diversity, great light, rapid changes, both social and environmental, and complex interactions between man and nature. The National Trust supported my continued work on the coast, resulting in this book.

For years I've challenged the separating out of the genres that we've inherited from painting, which has often excised mankind from the rest of nature. At the end of the 1990s I spent thirteen weeks in the Galapagos Islands for *National Geographic*, determined to show that the archipelago was not a pristine wilderness but a thriving human as well as natural phenomenon. In an earlier book, *The Time of Trees* (1999), I set out to show how trees were not merely objects of beauty or veneration but had clear economic and social ties to human welfare, that in turn accounted for much of their existence or lack of it in various parts of the world. The co-evolution of people and their environment is what's exciting. I had already spent time in the late 1970s and 1980s documenting social change around parts of the British coast and I envisaged these early journeys being included in a new journey with *Irini*, a 34-foot sloop that I'd sailed over from Sweden.

I decided to work with two camera formats—35mm to capture swift, passing moments, 6x6cm to capture stiller, more studied images and to render fine detail. In Britain we experience Atlantic depressions passing and wait for new ones to arrive. I came closest to capturing this at Formby Point. A nagging wind was blowing. A storm had passed across the Mersey Estuary, leaving a partially leaden sky sweeping over the dunes. And then, as if by magic, a tiny figure passed by under a voluminous sky, clutching sandals. On another occasion, in the middle of winter, I travelled to Tilbury Docks. I'd sailed up the Thames the previous winter and recalled an atmosphere reminiscent of Joseph Conrad's description of that estuary in *Heart of Darkness*: "A haze rested on the low shores that ran out to sea in vanishing flatness". I recall trying to resurrect this mood, armed with a Rolleiflex, and was lucky to find this perfect haze illuminating the discarded shopping trolleys in the docks where, during the 1950s and 1960s, so many new migrants from the Caribbean saw England for the first time.

Most authors setting out around the coast have stuck resolutely to the shore. Over the past few years I have sailed most of the way around the coast—either in my own boat or in other people's. Over the past few months I have sailed all along the even waters of the south coast, and then north as far as Oban in Scotland; through the churning waves off Land's End and the frightening loom of the Longships lighthouse; with the fast tides of the Bristol Channel; and beating into fierce seas off the Pembrokeshire coast and off South Stack on Anglesey. South Stack has two faces: the landsman's view is of a peaceful bird sanctuary; the pilot's warning to sailors describes a "strong and dangerous sea" that, had we not been sailing at night, would probably have looked quite terrifying. In the Irish Sea we encountered unexpected calm, with razorbills and gannets skimming the surface around us as we headed north from Larne.

The good spell lasted until we reached the Sound of Islay, where portentous thunder clouds were a prelude to a gustier sail through the Hebrides. Wind would change from dead calm to a near gale in a matter of seconds. So with the genoa downsized to a working jib we sped with the tide past the Islay distilleries, the raised beaches of western Jura and the disturbed water at the entrance to the notorious Gulf of Coryvreckan where lies one of the world's most dangerous whirlpools. Like the Maelstrom off the Lofoten Islands, the cruel Coryvreckan has claimed its share of victims; the writer George Orwell was fortunate to survive a boating accident there. Once in the lee of the islands the sea stayed manageably calm—at least until we turned east from the Mull of Galloway on the route south.

Seeing the British Isles from the sea is an unforgettable experience. Geography becomes graphic, as it does the first time one sees the Isle of Wight from, say, Southampton and remembers the Solent was, until 10,000 years ago, just a dry river bed. The shape of the land and its relationship to other lands becomes etched in the mind. Distances become tangible;

sailing up the North Channel to Rathlin Island, in one sweeping glance I could see the Antrim coast, the Isle of Man, the Mull of Kintyre and much of southwest Scotland.

Although our nation was once defined by seafaring and our earliest writings include the Anglo Saxon poem, *The Seafarer*, our enthusiasm for the coast began only in the eighteenth century with the spa towns. There began the notion of the seaside as a salubrious retreat for scrofula sufferers and an escape from the increasingly polluted cities. Before this time the sea was generally loathed and feared on both sides of the English Channel. Today, despite the twentieth-century revolution in air travel, our love affair with the British coast—our "sea fever"—takes us birdwatching, diving, dog-walking, fishing, sunbathing, surfing, sailing, coastal hiking or just looking out to sea. This book explores this jewel of an island's relationship with Shakespeare's "silver sea"—"whose rocky shore beats back the envious siege of watery Neptune."

Rugged cliffs, sea stacks, shingle spits and undulating dunes are all products, in their various ways, of geology meeting the power of waves and weather. Perched on the edge of the European plate, the British Isles has been jostled, bumped, buried, twisted, exposed, glaciated and shifted around for millions of years. Its mixture of rock types gives varied soils, and in turn a complex range of flora and fauna. And each time landforms have emerged they have been mercilessly battered by wind and rain. All this disruption and disturbance has given rise to a coastline unmatched, at least in terms of scenery, in Europe. It's a scenery the National Trust, through its 40-year Neptune campaign, has managed to snatch from the jaws of the developers, leaving many of our most beautiful beaches untouched by commerce. The Mediterranean coast may be hotter but the landscape has little geological diversity. The Mediterranean's predominantly limestone shore was formed in the Cretaceous era—a small window of geological time between 67–150 million years ago. By contrast Britain's rocky shores have a history of landscape formation going back more than 600 million years.

ON BOARD *IRINI*, PADSTOW, CORNWALL, 2005

Everything changes, especially where the sea is concerned. One of the main concerns of this book has been to try to understand that change from a variety of perspectives. Several themes kept cropping up again and again in my journey: the growing impact of climate change; the wider environmental problems brought by fishing and pollution (including litter); economic change, mostly through the repositioning of our manufacturing base; and social change—a result of ever more urban demands upon our coastal environment.

Our climate is changing. This is nothing new, but what is new is the fact, recognized officially in 1995, that *our role* in increasing concentrations of greenhouse gases, such as carbon dioxide and methane, has had "*a discernible influence*" on our climate. This has been happening for some time, and no matter what we do, will keep altering our climate and weather for the foreseeable future. There is nothing straightforward or uniform about this change. It is affecting many regions of the earth differently. However, the overall trend is that the climate is becoming warmer: thermal expansion and the melting of grounded ice sheets mean that sea levels are rising everywhere.

The warming trend is more marked on continental shelves, such as in the Baltic Sea or the North Sea than, say, the North Atlantic Ocean because on shelves the water is shallower. For example, the southern North Sea temperature has increased by 2°C over the past 20 years, far more than the increase on land. Other well documented trends linked to climate change in the British Isles are changing rainfall patterns and increasing storm activity. Coasts erode fastest during storm events; however, erosion is not always undesirable.

Erosion has sculpted our coastline. It has formed the iconic landforms we know well: the white cliffs of Dover, the Needles, and Dorset's Jurassic coast. No erosion would mean no dramatic, chalk-white cliffs. Yet frequent storms and increasing winter rainfall are speeding up erosion around the coast—and the pace of change is forcing communities and policymakers to choose, more than ever before, between engineering sea defences or letting nature take its course. "Making space for water" and "managed retreat" are recent policy objectives set out by both the government and the National Trust.

The emerging idea is to allow the sea to breach its long-standing natural defences, such as sand dunes and shingle banks, and flow freely inland. This allows the land itself to act like a "sponge" for the huge energy of the sea, mopping up the waves—stopping the monster from ravaging further down the coast. This is now occurring in several parts of the country: in Porlock, Formby, the Birling Gap and Happisburgh.

The benefits include creating new types of natural habitat such as salt-marsh. But people, by contrast, feel threatened. Many urban centres are exposed to flooding, such as London and Bristol. Rural communities liable to storm surges can be even more vulnerable, such as the crofters of the Outer Hebrides. In neither case do people feel inclined to retreat from the sea; on the contrary significant interventions, such as the Thames Barrier, continue to be built.

The post-war incentive for engineered sea defences around our coast was prompted by one terrible storm off the Norfolk coast on 31st January 1953. Overnight the pressure dropped to 966 millibars. The seas rose, then 125 mph winds drove the surging waves ashore, causing the loss of 2,107 lives in Britain and the Netherlands. Nobody then was interested in managed retreat. After all, where would it end? Resolve and defence for the sake of those who died were the imperatives of the day. The very same sentiments were impressed upon me whilst visiting Uist in the Outer Hebrides, where five people perished on 11th January 2005. Pressure was lower than in the 1953 storm—944 millibars; winds were just as strong or stronger and blowing onshore; and sea levels were higher—this being one of the added risks associated with global warming.

Six months after the January storm the sturdy and bearded figure of Murdo Mackenzie, a retired geography teacher, clambered onto a shingle bank to the west of Loch Altabrug on South Uist—the only protection the twenty or so crofting families have from the next Atlantic gale. He pointed to the stripey, blue-grey pebbles of Lewisian gneiss piled up beneath him and said simply: "it might as well be a wall of confetti." Nevertheless, when I suggested retreat he was adamant: "we couldn't retreat; there's nowhere for us to retreat to."

It was also one of Britain's worst rainstorms in living memory

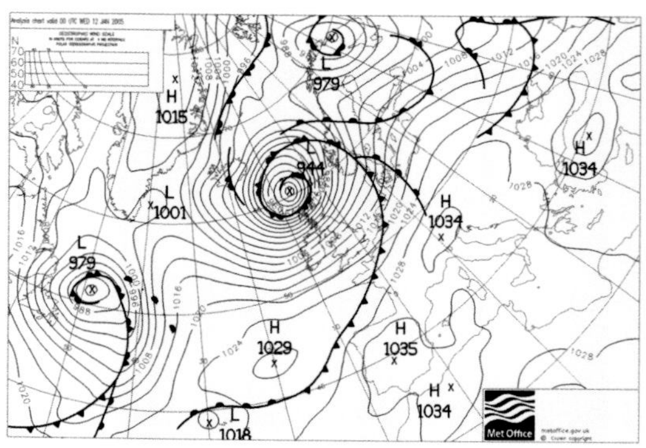

MET OFFICE WEATHER MAP SHOWING VERY LOW PRESSURE, 11TH JANUARY 2005

on 16th August 2004 that nearly destroyed the small town of Boscastle. Some will argue that there have always been catastrophic coastal storms: others point to climate change, yet again. Recall, for example, the two day hurricane in October 1859. It sunk 133 ships around the coast with the loss of 800 lives, including the gold-laden *Royal Charter*, wrecked off Anglesey. The hurricane off the south Cornish coast in December 1981 brought in its wake the Penlee lifeboat disaster, and bore witness once again to the bravery of perhaps the noblest of all British institutions, the Royal National Lifeboat Institution. The RNLI was founded in 1824 as the National Institution for the Preservation of Life from Shipwreck. Its motto, engraved on the reverse of the RNLI gold medal, is adapted from Psalm 69: "Let not the waterflood overflow me, neither let the deep swallow me up, and let not the pit shut her mouth upon me."

Fortunately, the same atmospheric physicists that have monitored changes to our climate are today able, with more accuracy than ever before, to track our weather. Countless maritime tragedies are avoided each year through HM Coastguard working with chartered meteorologists to follow weather systems from coastal stations, as well as geostationary satellites. On my travels around the coast I have asked frequent questions of lifeboat crews, fishermen, archivists, sailors and conservation officials about the change taking place around the coast. All recognize differing wind patterns and are deeply concerned about how this may affect their communities.

On the island of Islay archivist Dr Margaret Storrie has kept a diary since she was five. Until recently, there was never any mention of easterly winds. For the most part the prevailing wind came from the southwest. Nowadays winds are more easterly, ferocious, unpredictable, and ferries are frequently diverted. The National Trust wardens at Formby Point, north of Liverpool had also noted changing weather patterns, bringing persistent north-westerly winds that are eroding the fragile dune system—home to the rare natterjack toad. The collapse of the lifeboat slip at Happisburgh in Norfolk in the winter of 2002 seemed to offer clear evidence of worsening weather, although many blame Happisburgh's heart-breaking erosion problems on offshore dredging further up the coast.

Birds, fish, reptiles and mammals are also struggling to keep up with the pace of change. In 2004 there was catastrophic failure of seabird breeding, especially on Scotland's northeast coast. Derek Shaw of the Bird Observatory in Fair Isle was this year hoping for some improvement. Failure has been blamed on the impact of sea temperatures on sand eels, the staple diet of kittiwakes, puffins and guillemots.

The wider environmental impacts associated with the coast, such as pollution and over-fishing, have been another concern. Sewage outfall is a threat, but the continuing pollution of the coastal environment is especially visible in the form of litter washed up on our beaches. There are few beaches that are not habitually strewn with plastic containers, a problem that is not simply undesirable, but against international law. The Marpol Treaty, an international convention "providing for a cleaner, safer, marine environment"—a convention that we and most nations have signed—clearly

RNLI GOLD MEDAL

states that: "[it is] illegal for any vessel to dump plastic garbage including synthetic ropes, fishing nets and plastic garbage bags anywhere in the oceans or navigable water."

But dumping at sea is only part of the problem. Industrial or agricultural plastic sheeting is frequently blown off the land and out to sea, which is also a hazard to shipping, as I discovered to my cost. In a dead calm off the Mull of Galloway my boat's engine seized up and I became immobilized when a very large sheet of black agricultural plastic wrapped itself around the propeller. Happily, Portpatrick's lifeboat towed me into harbour where I could dry the boat out against the harbour wall and fix the problem.

The Royal Commission report on the impact of fisheries, *Turning the Tide*, was published in December 2004 shortly after Charles Clover's outraged book on "the scandal of overfishing", *The End of the Line*. The message at the end of 2004 rang out loud and clear: there must be radical changes in fishing practice if we are to protect our marine environment from total decimation. The Commission called for a statutory system of spatial planning that covers all major uses of the sea, including fishing. This would involve the closing of one third of our fishing grounds to commercial fishing. The concern is linked to a terrifying global assessment of human impact at sea. Over 90% of larger predatory fish may have been lost since the pre-industrial era.

Worse, it is now apparent that the damage caused by fishing is not confined to commercial fish populations, but affects the wider marine ecosystem. Deep water trawling, in fact most forms of trawling and dredging, for example for scallops, are damaging to the seabed, and especially so to coldwater corals and seamounts. In *The End of the Line*, Clover rails against the way industrial fishing practices are flouting the law in order to make an increasingly tough industry profitable. At close he names a dozen "fish to avoid"—which happen to be many of the fish easily available on today's seafood menus: among them Atlantic cod, haddock, Bluefin tuna, caviar, hake, sea bass, halibut and scallops.

Whilst working on this book I noted several changes to fishing and consumer practice in Britain. First, there seemed to have been, over the past 15 years, a geographical shift in activity away from the major east coast fishing ports of Grimsby, Hull and Lowestoft to the comparatively active ports of Brixham in Devon and Newlyn in Cornwall.

From the 1950s well into the 1980s, Lowestoft—reliant upon the Dogger Bank—was the principal supplier of plaice and sole to Billingsgate, London's premier fish market. Today the fishing port is virtually derelict. The south west coast meanwhile, though still active, could not be described as booming. Newlyn is home to small-scale trawlers or converted trawlers, such as David Pascoe's 32-foot *Little Pearl* that, in the winter of 2004/5, was netting abundant stocks of pollock which cluster close to the profusion of wrecks off the Cornish coast. British fishermen habitually complain either about the price they achieve at market or about the fleets of foreign or relocated boats destroying their business, whilst from their perspective their whole industry seems vulnerable, if not from challenges to the business then from government regulation.

Today the beam trawler fleet at Brixham is under threat. In just five years the number of trawlers in commission has decreased from 47 to 17 and the future is uncertain. One difficulty for British fishermen to manage is the fluctuating price of fish, an effect created by the near-global sourcing of fish by the market and by the profit extracted from fishing by traders and restaurant owners. In Padstow, Cornwall lobster fishermen complain about the difference between the price they command for lobsters—£3 to £4 each—and the £36 price tag for lobster on the menu of Padstow's upmarket seafood restaurants after boiling in a little water.

To some degree, fishing forms a social safety net for coastal or island populations without jobs onshore. Those who lose their jobs in the Islay distilleries have little choice (apart from relocating) but to go fishing. As radical steps are taken, such as closing off a third of our fishing grounds to commercial fishing, it is worth considering the broader employment nexus that the fishing industry occupies with a view to creating alternatives for those fishermen who stand to lose everything from such policy shifts.

The downturn in fishing is echoed around the coast in the virtual collapse, over the past 30 years, of our shipbuilding industry. Throughout the 1980s I documented the closing of shipyards on Teeside and Tyneside. I recall a small knot of men and women gathered in the sunshine in Smith Dock to watch the launch, in 1986, of the last ship to be built in the Tees estuary. On Tyneside I saw the shipyards close one by one. Men walked out, large and bitter gatherings took place and demolition gangs levelled the ground. "We are fighting being made into a museum." These words were not spoken on the Tyne, but on the river Clyde in 2005, by a naval architect named Ken Fulford. Fulford works for Ferguson's, the last shipbuilding firm remaining on the lower Clyde. Once Ferguson's employed 400 people, now they employ 250 and their future depends upon new orders.

The day I interviewed Ken Fulford he was competing fiercely with a shipyard in Gdansk, Poland, for the contract to build a British fisheries protection vessel that would keep his shipyard going. He was also preparing for the launch of their latest "ship"—a chain ferry for the Tamar river. What a change has happened in Port Glasgow, the home of British shipbuilding since 1711, when the Scott family of Greenock opened for business. In its heyday in the 1970s the shipyards around Port Glasgow employed 10,000 people, 4,000 of them at Lithgow's where some of the world's biggest ships were built. Great supertankers such as the *Nordic Clansman* and her sister ship the *Nordic Commander*, each about 800 foot long, were so large that they had to be assembled in sections using goliath cranes. This was before the skilled Scottish shipbuilders contracted to train the South Koreans how to build their own ships and the government chose not to subsidize manufacturing—and the rest is history.

The demise of our manufacturing productivity, at its lowest now since records began, shifts in fishing practice, and continuing urbanization around the coast has led to changes both to the built environment of our coastline and our relationship to the sea. These changes are most noticeable in coastal cities such as Newcastle or Liverpool. Along the river Tyne, Newcastle's waterfront has been reclaimed by developers. Expensive flats, lawyer's offices, a flour mill-turned-art gallery, have replaced the derelict industrial clutter of the inner city wharves.

At Wallsend the cranes still stand. Little of Hadrian's wallsend is visible, much less the fort. Most of the stones were sold off for building long ago or pilfered by monks from Tynemouth. But there is income to be generated from school trips to the Roman bath-house, a local history museum, the viewing tower, and the café where "Hadrian" mineral water, sparkling or still, is on sale. "Heritage-led regeneration" is developer-jargon for the commodification of the past in order to build the future. In Britain, Liverpool is the prime example of this

BREAKING UP THE TYNE SHIPYARDS, 1986

trend. Over seven miles of the Mersey Liverpool has what John Hinchliffe—World Heritage officer for the city—describes as "the biggest and best preserved system of historic docks anywhere in the world." Today the city's mercantile past has been rewarded with World Heritage status, and cultural tourism has become the pioneer industry of the city.

Cheap flights, increased leisure time and more disposable income have allowed cities like Liverpool to base their future prosperity on people walking around gazing at things and shopping rather than being active "doers". The city already boasts the Tate Liverpool, part of the substantial Albert Dock development, the Beatles Museum and the Maritime Museum. To come is the King's Dock development with a 9,000-seat arena and conference centre, a "boutique hotel" and, nearby, a new museum—the Museum of Liverpool—to accompany the Museum of Liverpool Life. The Paradise Project is a retail development park to bridge the gap between the docks and the city centre with a million square foot of shopping opportunities. The whole scheme is due for completion in 2008 when Liverpool becomes European Capital of Culture.

Where the large city has led, smaller cities, towns and regions have followed. The Fal estuary is an example. The aim of the National Trust there is to acknowledge the importance of tourism but at the same time to try to reduce its environmental impact—through the green transport initiative. Tourism is transforming the social fabric of the coast, where the service revolution has replaced the industrial one. Perhaps half of today's coastal population consists of either tourists or the people whose job it is to serve them. Temporary, part-time or seasonal jobs involved in, say, manning car parks, cleaning hotel rooms, or selling bottles of water or ferry tickets are ten a penny: secure or well paid jobs are scarce. Second homeowners raise house prices over and above the earning capacity of the locals, driving them away.

In this voyage of discovery I hiked and sailed through so many impressive places: walking the glorious coastal footpaths of Cornwall and Pembrokeshire, hiking to Warbarrow on the Dorset coast, the Needles, sailing through the Narrows to Strangford Lough and visiting several of the "pladdies"—very large rocks that become islands at low water; and the magical Scottish islands of Lunga, Ailsa Craig, Eirrad, Canna, Muck, the Ross of Mull and Easdale—queen of the Slate Islands.

One thing about the British Isles, that I think only black and white photography can truly capture, is the quality of light. It changes constantly. I have encountered wonderful light in other parts of world, from the diffuse, ethereal light of Costa Rica's cloud forest to the warm evening light of the Cycladic islands in Greece. But here, in places like the Menai Straits, the Small Isles of the Inner Hebrides, and the Farne Islands, it is the ongoing contest between the rain and the sun, mirrored in the sky, that is unique and to which the entire landscape—human and natural—conforms.

Mostly though I recall the people whose lives *are* the sea. People such as Willie Ramsay, Portpatrick's harbourmaster; the MacEwens of Muck; Commodore Angus Sandford, with whom I travelled up the Clyde; the crews of the Clyde Marine Unit; Murdo MacKenzie and the McRurys of South Uist; Neil Brockman—coxswain of the Penlee lifeboat; Tony Barclay and Rod Pace of the Moelfre lifeboat; Lieutenant Commander Ahmed Ajala in the Solent—and National Trust wardens, managers and volunteers too numerous to list here. So many of the people I met on these journeys around Britain's coast had a humility, grace and generosity of spirit. I wondered why. Perhaps because the ever present spectre of the sea, cruel and unpredictable, makes such men and women more keenly aware of a shared vulnerability.

The book's title was chosen because I wanted to draw attention to our coastline, and the enjoyment we all get from just thinking about it. Coastal pathways have opened up the pleasures of walking by the sea to a wide spectrum of society. More people than ever are taking to sailing, surfing, diving and windsurfing. As people take more time off or have more leisure time, their preferred destination is often "beside the seaside". Whether this means Pride Week revellers with artful costumes heading for Brighton, surfers tapping into dedicated websites in search of the best waves, City workers headed for Cowes, walkers planning a long walk along the Cornish coast, or day-trippers off to Inner Farne or one of the other 400 or more islands off our coast, our coastline is certainly enjoyed as never before.

postscript Although most of the photographs in this book were taken over the past year and a half, my relationship with the British coast really began in 1979. As a young student I joined the container ship *Atlantic Conveyor* on a voyage from Southampton across the Atlantic to Baltimore, to document the changing face of merchant shipping in the aftermath of containerization. I travelled with Captain Ian North DSC, who was kind enough to accommodate me in the owner's suite on the ship and to teach me, over several weeks, so much about the sea. It was with great sadness that I learnt of the tragedy that befell the *Atlantic Conveyor* and Captain North during the Falklands Conflict in 1982. The ship was hit by two Exocet missiles in the afternoon of 25th May and eventually sunk. *Atlantic Conveyor* was defenceless against heat-seeking warheads and ill-prepared to deal with their highly flammable propellant. Only a month earlier I'd been in Southampton dock to watch the aircraft carrier *HMS Invincible*, its sailors saluting proudly on deck, as it departed for the southern ocean. The naval engagement in the Falklands was remarkable for its brevity, its intensity and for its tales of the unexpected. As in war, so in our experience of the forces of nature *sudden and unexpected change* will continue to shape our coastal community and environment.

HMS INVINCIBLE LEAVING FOR THE FALKLANDS, SOUTHAMPTON, 1982

ARCTIC TERNS, BROWNSMAN ISLAND, FARNE ISLANDS, 2004

farne islands, june 04 Before Alfred Hitchcock's 1963 film *The Birds* even begins, dark gull-like silhouettes are attacking and destroying the credit-lines as they roll up on screen. Cut to the Farne Islands. I'm walking the gauntlet of Arctic terns en route to the small warden's cottage on Brownsman Island. Before we even begin working it feels as if I've been pecked to death. Why the birds lay their eggs right on the path outside the cottage is a mystery, but having done so they cannot be blamed for using every scare tactic in the bird book to keep intruders away. But the resemblance to a horror film ends there. I donned a builder's hard hat whilst on land and spent an enjoyable three days watching the redoubtable wardens—Alex Ash, Neil Dawson, Dave Clare and Rob Wheeldon—counting sea birds and checking their nests. I never really understood how they added up some 40,000 guillemots plus kittiwakes and razorbills whilst bobbing about on a rubber boat in a swell at six in the morning. That's why I photographed the wardens as heroes, statues carved out of the morning sky. The Farnes was the second place where I'd been advised to wear a hard hat on the beach. The other, a few months earlier, was when I was following geography students from Newcastle University learning how to describe glacial deposits at Whitley Bay. I have an indelible picture in my mind of twenty or so students, slimline versions of Bob the Builder, in yellow hard hats, marching through the marram grass clutching micrometers and geological hammers.

scarborough, july 05 It's early morning. Fog is thick on the beach, over the Italianate villas and the Grand Hotel. Three hundred years ago this same beach, with its ugly trickle of spa water and honey-coloured sandstone walls, had woken up to find itself the world's first seaside resort at the beginning of a cultural revolution: the sea was no longer to be feared, loathed or seen as a barrier. Instead the sea came to represent, as the historian John Walton put it, a cure-all for the idle, the under-exercised and the overfed. First to arrive on the beach today are Vince from Wales and his friend Mary. They are the jackdaws of the sands, searching for coins, rings or anything of value with metal detectors. It's 8am and already Vince has scooped up about £1.80 in change. Once he got lucky and found a diamond ring. That was in Barmouth. Next to arrive are Nick Granville-Hall, a music student from Leeds, with six donkeys and two helpers, Lauren and Karen. They charge £2 a ride: one child wouldn't go home until he'd been on all six donkeys.

skegness, july 05 As dark clouds gathered, a lady walked into the tourist information kiosk on the esplanade, protesting: "The weather forecast said it was supposed to be sunny." "I think it is, probably," replied the official, "above the clouds." A short while later the skies opened. Weekenders donned absurd bright orange ponchos, bought umbrellas from well prepared traders or sought shelter in the many bars. There's an understandable obsession with the weather in Skegness; but at the same time a blithe optimism. The manager of the Quality Rock Shop, dressed in a pink-striped pinafore, enthused: "It's always sunny in Skegness. It's just that the clouds get in the way—we try to be jolly about it." Such stoicism probably led to the "jolly fisherman's" famous slogan: "Skegness is SO bracing", in a word converting the cold and miserable northerly winds that lash this coast to something beneficial and salubrious. The slogan was used by the Great Northern Railway to advertise excursions from London's Kings Cross station as early as 1908.

blakeney point, november 04 The best place to cycle around the British coast, if the thought of steep hills and aggressive motorists is too much, is Norfolk. It's flat and they even have a royal road just for cyclists—National Cycle Network Route One. The route passes the Sandringham estate then meanders down the most idyllic, tree-lined cycle track in Britain into King's Lynn. I had travelled in this way from Great Yarmouth, mostly in the pouring rain, aiming to see the extent of coastal erosion at Happisburgh and to visit Blakeney Point which by contrast is not eroding but growing. The strange thing about cycling or sailing (as opposed to driving) is that although travelling much slower than by car—somewhere between four and fifteen miles per hour—I had the sense of seeing much more. Also, I seemed to be making better pictures this way. Because I was already outside and often soaking wet it was no extra effort to explore another beach or bay or village.

the solent, july 05 Naval efficiency ensured that the Solent Fleet Review—an inspection of 167 vessels—ran smoothly for all involved. That was Commander James Morley's main concern as he stood on the bridge-wing of the frigate *HMS Lancaster.* Especially when the Queen passed by on the ice patrol ship *HMS Endurance.* "Hurrah!" shouted the mustered ship's company as caps were ceremonially removed. On every ship in sight men and women were positioned formally on deck. Submariners stood along the hull; Colombian sailors manned the cross-trees of a tall ship nearby; and French pilots and crew looked down from the flight deck of the carrier *Charles de Gaulle.* For Morley the most significant aspect of the event was that 43 of the 167 vessels taking part were not British. The term "naval diplomacy" (once referred to as "gunboat diplomacy") has taken on new meaning over recent years. The man who explained this to me most lucidly was Captain Russell Best OBE whilst aboard the frigate *HMS Cumberland* on the Clyde. Best sees the Navy's role "as a more general arm of government abroad". The Navy's "best of British" brand of diplomacy is an important factor in keeping the peace, in securing contracts overseas, in bolstering trade and acting as a "floating embassy": especially valuable in regions such as south-east Asia. For this reason many senior naval personnel regret the decommissioning, in 1997, of the Royal Yacht *Britannia*, a ship that in its day sailed to over 600 ports in 125 countries.

cowes, august 05 "Let's make babies" was the boat's motto but given the conditions at sea and the racing agenda, this would have been a challenge. Neil Peters, the affable owner of *Baby and Toddler Gear* magazine, had invited me racing with his expert crew aboard *Baby Gear* on a blustery day during the annual Skandia Cowes Week. We were in the Sunfast 37 class, the corporate hospitality set, pitted against a variety of high street banks and other finance firms. Tina Sexton, a business development manager, was navigating whilst Tom Gall, a nineteen-year-old recruited from a local sail shop, demonstrated how to sail downwind. Our figurehead comprised two soaking wet baby dolls tied to the forestay. It was an exhilarating day, especially for the spinnaker, which at one moment was flying from up top like a flag and another moment was being recovered from the sea. I'm not sure where we came in the race; that's not the point. The point is that we were out racing with 941 other boats that day, the best year ever in the biggest week-long regatta in the world.

ARCTIC TERNS, BROWNSMAN ISLAND, FARNE ISLANDS, 2004

NATIONAL TRUST WARDENS, BROWNSMAN ISLAND, FARNE ISLANDS, 2004

BIRD SANCTUARY, NEWTON LINKS, NORTHUMBERLAND, 2004

NATIONAL TRUST WARDEN, BROWNSMAN ISLAND, FARNE ISLANDS, 2004

NEIL DAWSON AND DAVE CLARE, NATIONAL TRUST WARDENS, FARNE ISLANDS, 2004

DAVE CLARE AND ALEX ASH, NATIONAL TRUST WARDENS, FARNE ISLANDS, 2004

TOURISTS MIX WITH SHAGS ON STAPLE ISLAND, FARNE ISLANDS, 2004

EMBLETON BAY, DUNSTANBURGH CASTLE, NORTHUMBERLAND, 2004

EASINGTON, CO. DURHAM, 2000

LINDISFARNE, NORTHUMBERLAND, 2004

FOOTPATH LEADING TO DUNSTANBURGH CASTLE, NORTHUMBERLAND, 2004

THE LAST SHIP TO BE BUILT ON THE RIVER TEES, MIDDLESBOROUGH, 1986

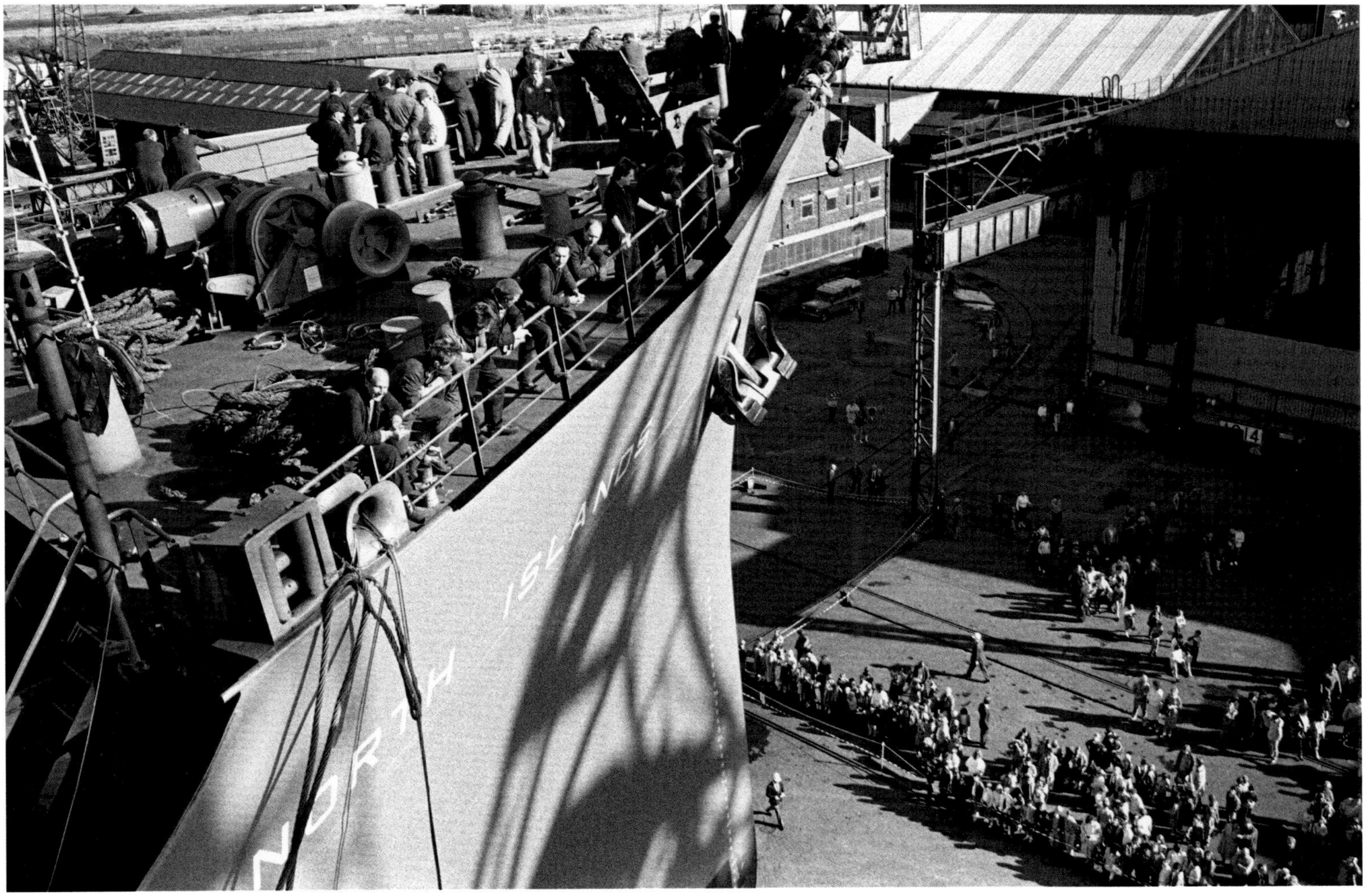

THE LAST SHIP TO BE BUILT ON THE RIVER TEES, MIDDLESBOROUGH, 1986

DIVER, BRAE OIL FIELD, NORTH SEA, 1985

PIPE DECK, BRAE A PLATFORM, NORTH SEA, 1985

BREAKING UP THE TYNE SHIPYARDS, 1986

HERITAGE CENTRE, WALLSEND, TYNESIDE, 2002

CHIPS WITH KETCHUP, FILEY, YORKSHIRE, 2005

PLEASURE BEACH, SKEGNESS, LINCOLNSHIRE, 2005

NEWTON BEACH, NORTHUMBERLAND, 2004

WINTERTON-ON-SEA, NORFOLK, 2004

DRIFTWOOD, WINTERTON-ON-SEA, NORFOLK, 2004

BAKERY, SKEGNESS, LINCOLNSHIRE, 2005

DONKEYS RIDES, SCARBOROUGH, YORKSHIRE, 2005

TIMBER REVETMENTS, HAPPISBURGH, NORFOLK, 2004

WHITLEY BAY, NORTHUMBERLAND, 2002

THE SEA FRONT IN SHERINGHAM, NORFOLK, 2004

VISITING THE SEAL COLONY AT BLAKENEY POINT, NORFOLK, 2004

SEALS AT BLAKENEY POINT, NORFOLK, 2004

TILBURY DOCKS, LONDON, 2004

TILBURY DOCKS, LONDON, 2004

LIFE ON THE *ATLANTIC CONVEYOR*, NORTH SEA, 1979

PRIDE WEEK, BRIGHTON, 2005

PRIDE WEEK, BRIGHTON, 2005

FESTIVAL OF THE SEA, PORTSMOUTH, 2005

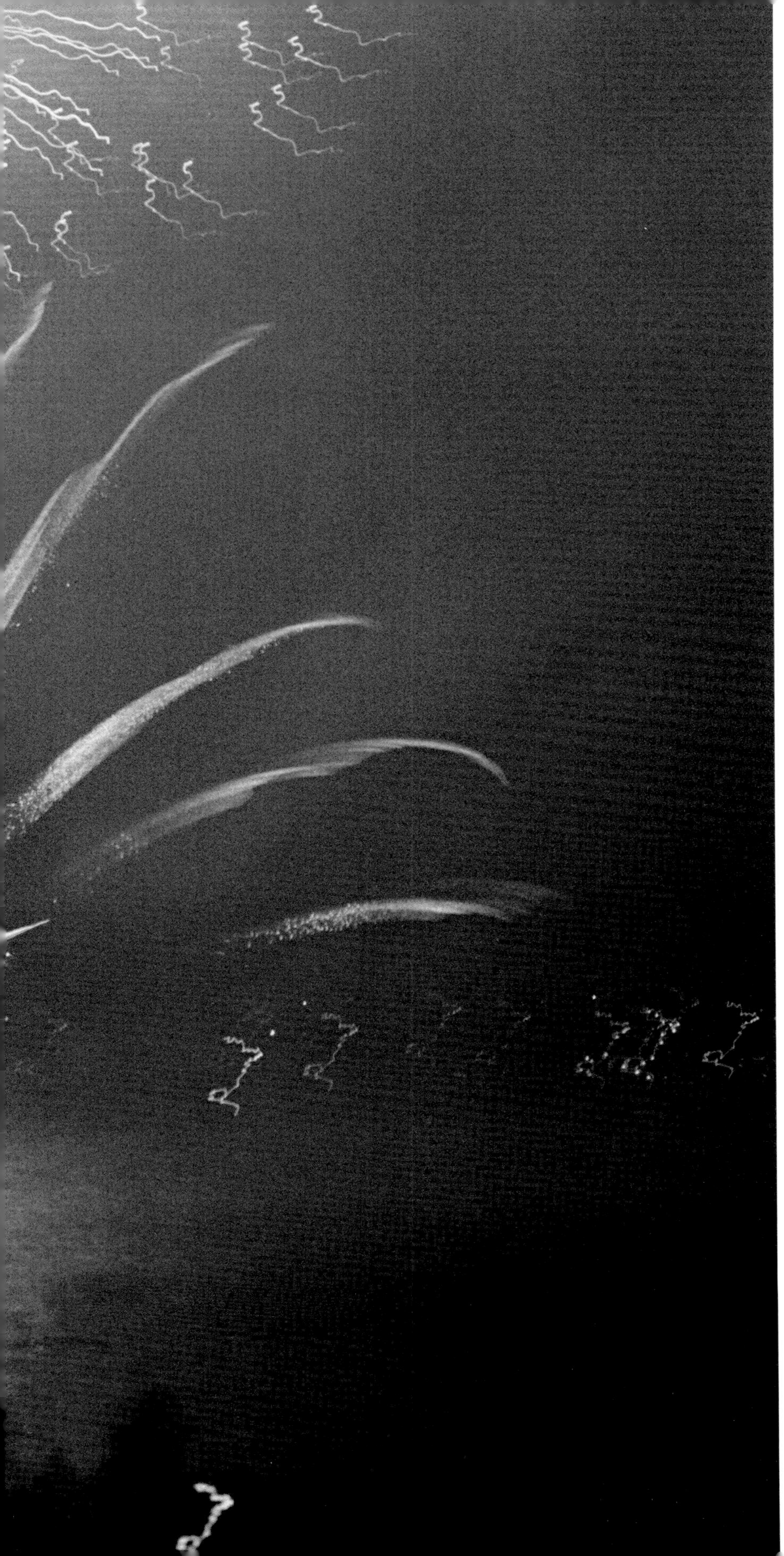

RE-ENACTMENT OF THE BATTLE OF TRAFALGAR, SOUTHSEA, 2005

HMS LANCASTER, SOLENT FLEET REVIEW, PORTSMOUTH, 2005

HMS LANCASTER, SOLENT FLEET REVIEW, PORTSMOUTH, 2005

HMS LANCASTER, SOLENT FLEET REVIEW, PORTSMOUTH, 2005

HMS LANCASTER, SOLENT FLEET REVIEW, PORTSMOUTH, 2005

CAPTAIN IAN NORTH DSC,
ON THE *ATLANTIC CONVEYOR*, 1979.

RACING *BABY GEAR* AT COWES WEEK, 2005

RACING *BABY GEAR* AT COWES WEEK, 2005

BRONZE STATUE OF RICHARD EVANS, MOELFRE, WALES, 2005

the needles, july 05 Just after dawn bright white shapes like icebergs emerge from a blanket of mist. These are the Needles, tooth-like chalk stacks protruding into the English Channel. The name "Needles" is believed to have derived from a tapering rock pinnacle known as "Lot's Wife" which used to stand a little to the north of the present central rock. This needle-shaped rock, about 120ft high, collapsed noisily into the sea in 1764. Its stump can still be seen at low water, where it forms a dangerous reef. The previous evening I visited Alum bay, where fog spilt over the cliffs and small chalk deposits were encircled like jewels by the incoming tide. The undercliff's serenity and dawn light on the chalk cliffs has, unfortunately, to be set alongside the eyesore of the Needles Pleasure Park. The claim that half the Isle of Wight is protected as an Area of Outstanding Natural Beauty would be admirable were it true. Unfortunately, "heritage" developers have been permitted to place such features as "Jurassic Golf" and "Junior Driver", a children's car track with street signs, right in front of The Needles.

isle of purbeck, july 05 National Trust warden Emma Wright took me to see the Studland Bay naturist beach—the largest in the country. Walking up into the dunes behind the white sand beach we encountered men wearing nothing but binoculars, peering over the long grass like meerkats in the Masai Mara. Studland's naturist beach is a place of extremes: from well-intentioned men and women looking for a place to relax unclothed, to strangely motivated individuals with other ideas. Police "lesbian and gay liaison" officers were out amongst the nudists, leafleting, in an attempt to deal with the antisocial element. The National Trust's duties include protecting the dunes from the public and protecting the "textiles" (as the nudists call the non-naked public) from being the subject of "harassment, alarm, or distress." Police in earnest conversation with huddled groups of nudists begs the question: what were they discussing? Car parking, I was surprised to learn, was the most popular topic of the day.

dartmouth, april 05 A parade of ninety young officers was passing out, marking the culmination of the first part of their training. The event took place on the asphalt parade ground of the Britannia Royal Naval College overlooking the River Dart and the English Channel. Under the watchful figurehead of Britannia, they listened to the steady voice of the most senior person in the armed forces, General Sir Michael Walker. He said that it was an officer's duty to put those men and women who would be placed under their command before themselves; to value their lives above their own. Furthermore, they'd have to earn respect: it didn't come with the ceremonial sword and the first salute. Later, rain swept the parade ground as the Queen's Sword and Queen's Binoculars were handed out to those who had distinguished themselves whilst training. A naval officer's life must be one of contrasts, between the stiff, old-fashioned etiquette of the parade ground and off-duty informality. I caught a glimpse of the latter at the College's Centenary Ball. Indoors they danced the conga to "(Is this the way to) Amarillo". Out on the parade ground other officers navigated the contra-flow on a dodgem track with lunatic determination. Fortunately, on board ship, with such errant tendencies worked through, the Navy runs as it should.

lundy, april 05 In Julian Barnes's novel, *England, England*, it was suggested that Britain could be recreated as a theme park on the Isle of Wight. Seems things there have gone far enough in that direction. For quintessential England try Lundy. Good natured people, bent slightly, walking along tracks of stone and grass; wheeling seabirds; craggy and sublime granite cliffs; the sea all around; a coal fire; tea and biscuits; a book on smugglers; solid unfussy furniture; dry stone walls and kindly men who come to repair them. The only thing missing, for the moment, are puffins, the philatelic symbols of the island. There are only three. The problem has been egg-eating rats, which have recently been eradicated. Only twenty-three people live on Lundy. Among them is Kevin Welsh the farmer, who manages all the animals on the island, feral and domestic. It's the lambing season so by night he assists with the births and by day he drives around on a quad bike delivering feed. I took warden Ben Sampson's "rocky shore ramble". He pointed out cold water corals, sea anemones and tiny rock crustacea to a small group of visitors. Much time was spent on the mating habits of barnacles. Barnacles are among the most inventive love-makers on the planet; helped by some odd characteristics. If they were the size of humans, for instance, the male barnacle would possess a sixty-foot penis with which, if all else fails, it can mate with itself. At the end of the Lundy day all meet at the Marisco Tavern—a pub that never closes "in case there's a shipwreck".

milford haven, april 05 Reaching the Pembroke coast from Padstow was a thirteen hour sail. A strong south-westerly breeze took us 17 miles to the west of Lundy island where for an hour I photographed a pod of eight common dolphins on the bow. I was hanging off the boat and so low in the water that every time they surfaced I could feel their breath and their dorsal fins gently touching my feet. We made it to the lock gates in Milford Haven just a few minutes before they closed. Milford Haven is dominated by disused oil terminal jetties, but along a poorly marked coastal path out of town it's very pleasant, especially the beaches to the west of the bay, such as Sandy Haven. Hiking the Pembrokeshire coast is an utterly joyful experience and especially to see it in the good company of Pete and Gina Smithies and their three children. The Smithies farm 400 acres of National Trust coastline; much of the land is being converted back to coastal heathland.

moelfre, anglesey, may 05 Moelfre is a small inlet on the north coast of Anglesey, Wales. I went there especially to see the bronze statue of famous lifeboatman Richard Evans (1905–2001), who won two RNLI gold medals and a bronze medal. The sculpture shows Evans gripping the wheel, steadfastly steering the lifeboat into a rough sea, represented by rock. His most hair-raising rescue was of the eight crew of the Cardiff coaster *Hindlea* in huge, lashing seas off the Anglesey coast in October 1959. To rescue the crew Evans, a village butcher, had to take the lifeboat alongside the coaster ten times during one of the worst hurricanes ever to hit the Welsh coast. Of course Evans didn't work alone. His lifeboat crew, which included Donald Murley Francis, Hugh Owen and Hugh Jones, deserves as much credit and probably a statue as well. I put this question to the sculptor, Sam Holland. It turns out no piece of work has obsessed her so much. She spent two years modelling the clay figure (later cast in bronze) and worked every day for an entire month just on the eyes. In answer to my question about whether the whole crew should be honoured, Sam responded that the statue "stands for all lifeboatmen not just in Moelfre, but all over the country". One of those lifeboatmen was my father-in-law who was also at sea that night in 1959 aboard the Beaumaris lifeboat, sent to relieve Evans's boat as it went on to answer a second distress call. Evans's iconic status was duly recognized by Prince Charles, who came to unveil the sculpture in 2004.

THE NEEDLES, ISLE OF WIGHT, 2005

ALUM BAY, ISLE OF WIGHT, 2005

PASSING OUT PARADE, BRITANNIA ROYAL NAVAL COLLEGE, DARTMOUTH, 2005

THE RIVER DART, DARTMOUTH, 2005

CENTENARY BALL, BRITANNIA ROYAL NAVAL COLLEGE, DARTMOUTH, 2005

CENTENARY BALL, BRITANNIA ROYAL NAVAL COLLEGE, DARTMOUTH, 2005

BEN SAMPSON, NATIONAL TRUST WARDEN, LUNDY ISLAND, 2005

ROCK POOL, LUNDY ISLAND, 2005

KEVIN WELSH, LUNDY ISLAND, 2005

LUNDY LAMBS, LUNDY ISLAND, 2005

LUNDY ISLAND, 2005

TOM LOVEGROVE, LUNDY ISLAND, 2005

OLD HARRY ROCKS, DORSET, 2005

SCONCE POINT, ISLE OF WIGHT, 2005

POLICE LEAFLETING NATURISTS, STUDLAND BEACH, DORSET, 2005

NATURISTS ON STUDLAND BEACH, DORSET, 2005

COASTAL EROSION, PORLOCK, SOMERSET, 2005

SALTMARSH, PORLOCK, SOMERSET, 2005

RECOVERING AFTER STORM, BOSCASTLE, CORNWALL, 2004

STORM DAMAGE, BOSCASTLE, CORNWALL, 2004

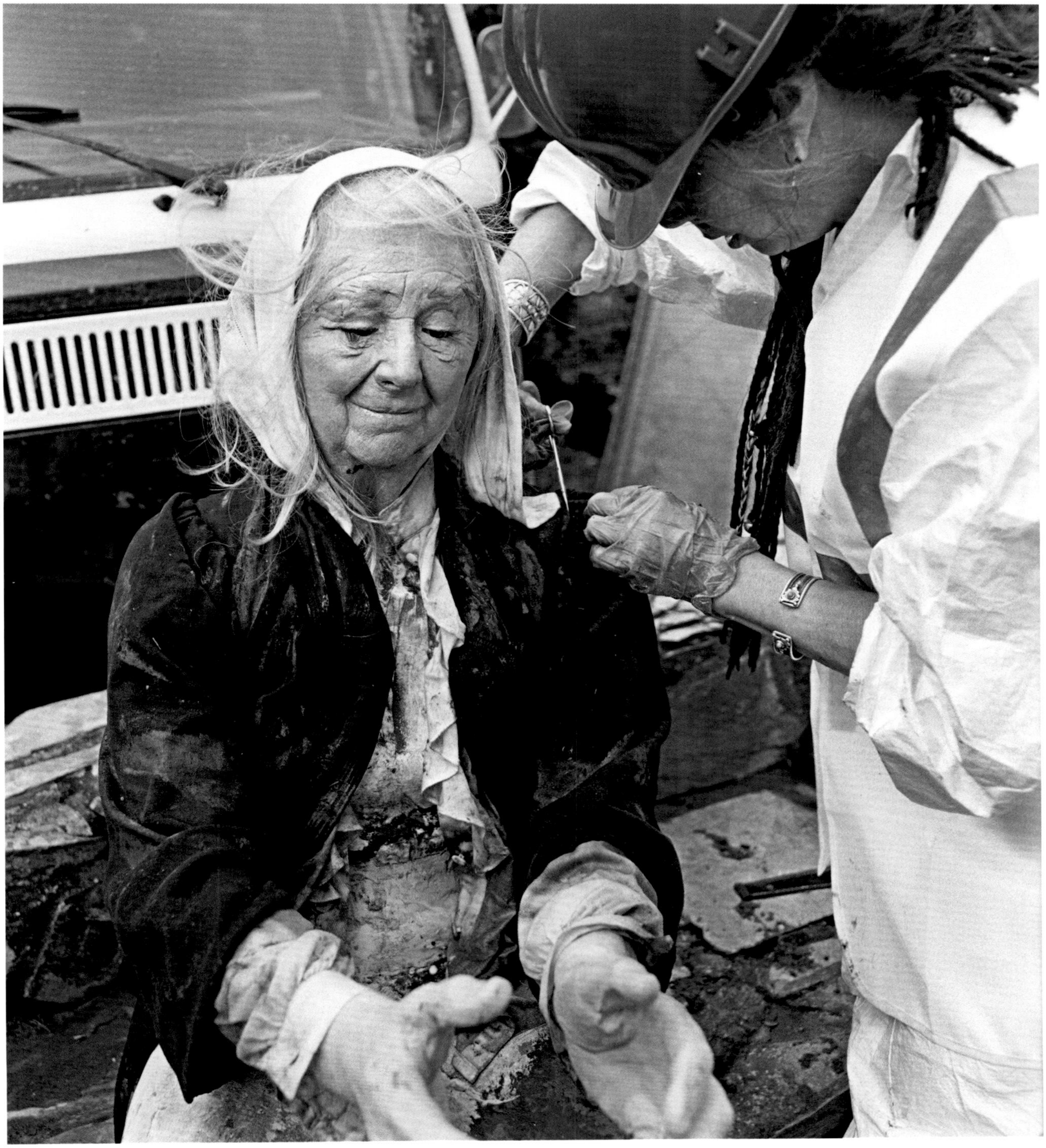

MUSEUM OF WITCHCRAFT AFTER STORM, BOSCASTLE, CORNWALL, 2004

BOSCASTLE, CORNWALL, 2004

JOHN BLACKFORD OYSTER FISHING, FAL ESTUARY, 2004

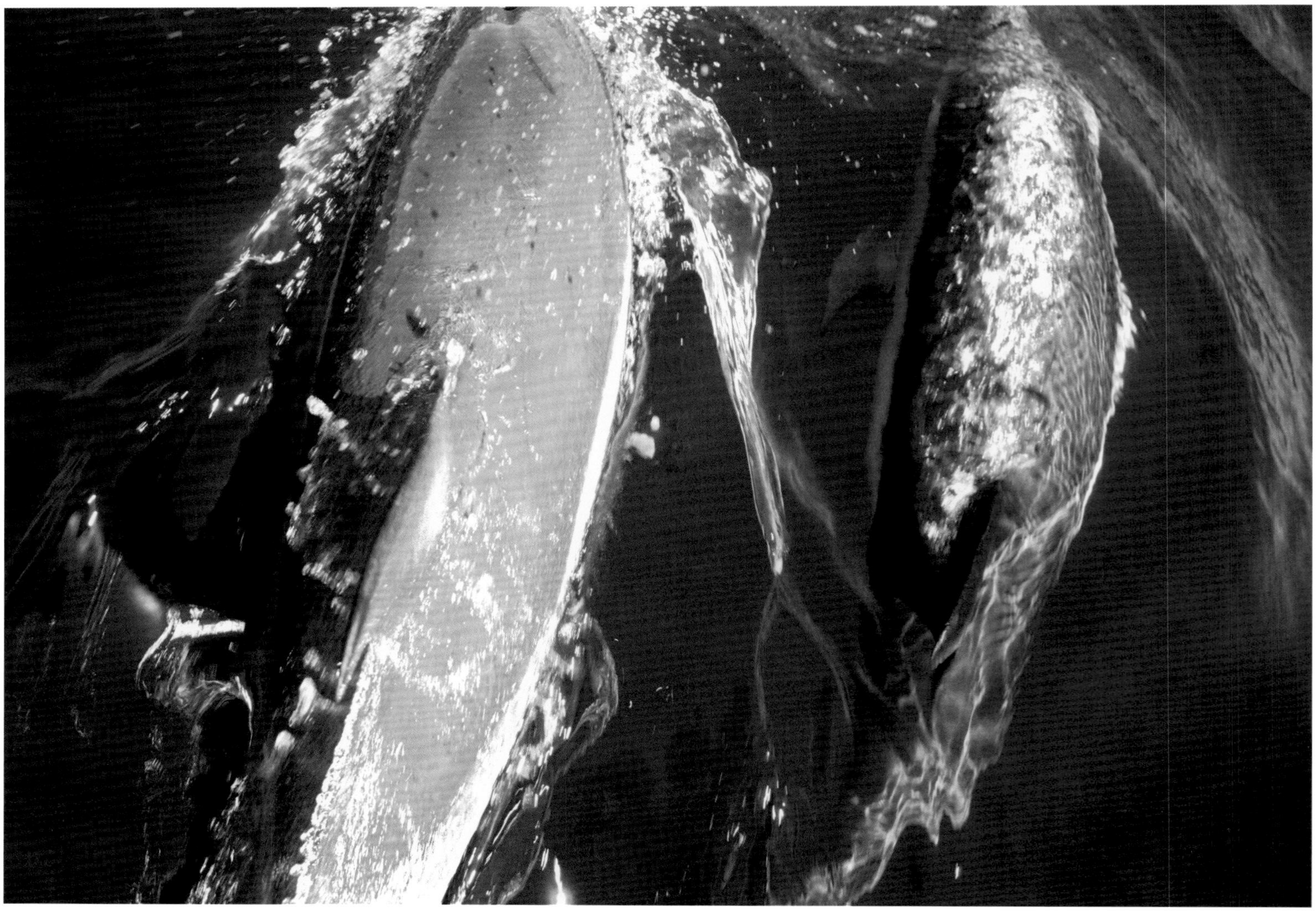

DOLPHINS, BRISTOL CHANNEL, 2005

SURFERS, PORTHLEVEN, CORNWALL, 2005

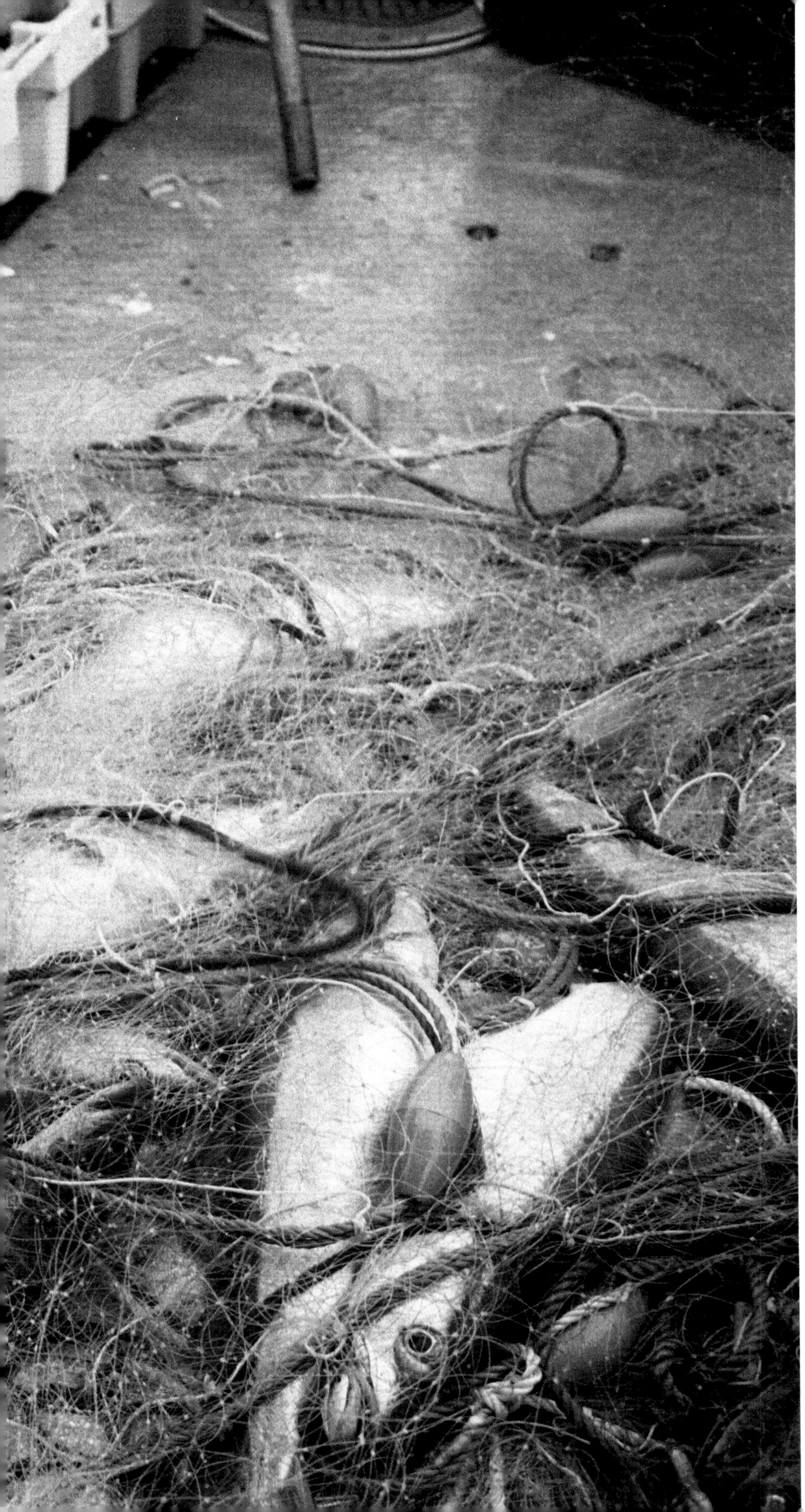

POLLOCK FISHING, NEWLYN,
CORNWALL, 2005

POLLOCK FISHING, NEWLYN, CORNWALL, 2005

ANTHONY BARCLAY, COXWAIN, MOELFRE LIFEBOAT, ANGLESEY, 2005

MARLOES SANDS, PEMBROKESHIRE, 2005

COASTAL FOOTPATH, MARLOES SANDS, PEMBROKESHIRE, 2005

SANDY HAVEN, PEMBROKESHIRE, 2005

PETER SMITHIES, MARLOES, PEMBROKESHIRE, 2005

NEIL ROTCHFORD, SPIDER CRAB FISHING, STACKPOLE QUAY, PEMBROKESHIRE, 2005

ICE CREAM VAN, TENBY, PEMBROKESHIRE, 2005

CARAVAN SITE, RHYLL, NORTH WALES, 2005

CAPE CORNWALL, 2005

MENAI STRAITS, NORTH WALES, 2004

MENAI STRAITS, NORTH WALES, 2004

FORMBY POINT, MERSEY ESTUARY, 2005

the ross of mull, may 05 Set out at 7am to walk across the Ross of Mull, over the bare hills of 400 million year old Caledonian granite; the same pinkish Mull granite that was used to build such London landmarks as Blackfriars Bridge, Holborn Viaduct and the Albert Memorial. As I pressed on through heather and creeping willow, cuckoos called, curlews flittered overhead and a mountain hare brushed past. The wave-like form of ancient farm plots (lazy beds or runrigs) was revealed on lowland, evidence of land once intensely cultivated before the late ninteteenth-century Clearances drove the crofters away. Today a small charity called Highland Renewal are bringing diversity back to a landscape once grazed flat by Soay sheep. The Ross of Mull is the nearest I've come in the British Isles to the sense of home-grown wilderness Thomas Hardy must have felt when writing *The Return of the Native*. There's no sound of cars and at night, apart from some small house-lights across the Sound of Iona, very little detracts from the brilliance of the night sky. But we are far from Hardy Country: this is Robert Louis Stevenson's domain. It's where the fictional David Balfour in *Kidnapped* was shipwrecked on the "treacherous granite fangs" of the Torran rocks, and washed up on the tidal island of Eirrad. It's where Stevenson himself sat in the garden shed of a lighthousekeeper's cottage writing fiction. The place is idyllic. My only problem came when attempting to return to Craignure. There were no buses so I hitchhiked a lift from a tree planter with two Patterdale terriers. "The dog's farted. Sorry." The occupant of the driver's seat wrestled one of the dogs off her lap. I had to help her, periodically pulling the black one into the back seat. "It's an ecological disaster", the driver was saying, pointing to the barren hills on the far side of Loch Scridain, where a combination of sheep and industry had denuded much of the island. I was still coping with the ecology of the car's interior.

fingal's cave and the treshnish islands, may 05 The tree planter of Mull wasn't the first to notice the missing trees. The first thought the burly New Zealand stone sculptor John Edgar expressed, whilst crossing from Mull to Staffa was: "where have all the trees gone?" It's true, there are hardly any trees left on the Hebrides compared to the way things were a few hundred years ago. The sculptor was collecting rocks for an exhibition in New Zealand. Today he'd come with his wife to see Fingal's Cave on the island of Staffa, a sculptural marvel in its own right carved from polygonal columns of basalt in the middle of the sea. About 40,000 people visit Staffa each year, mainly by launch from Mull. Ian the launch driver likes to liven the atmosphere with a recording of Mendelssohn's *Hebrides Overture*, written after the composer visited the island in 1829. From Staffa we visited Lunga, one of the Treshnish islands. I was travelling with a group of conservationists from Scottish Natural Heritage and the National Trust for Scotland. Lunga's one of those rare jewels that needs nobody's help. There are no rats, no people and the puffins are so tame you can virtually stroke them. Birders know all about Lunga, one of the finest places to look at seabirds in Europe, and they keep their knowledge very quiet.

muck, may 05 I first spotted the Laird of Muck in a stiff breeze rowing to shore in a blue plastic tub wearing phone-box red overalls. I recall something bright yellow as well: a sou'wester, sea boots? A man, in any event, of primary colours, Lawrence MacEwen has the gnarled look of an Edwardian explorer who wouldn't have looked out of place pasted onto a group photograph of Shackleton's Elephant Island expedition. He was bringing diesel drums from the mainland in a 35-foot coaster called *Wave*—without help. Conditions at sea were not good and I'd elected to leave *Irini* in Dunstaffnage about 50 miles down the coast. The anchorages on the Small Isles can be very exposed, not a problem for *Wave* tied up to the harbour's only mooring. Rough weather didn't trouble the Laird who strode into the Craft Shop, the island's sole café, greeted his wife Jenny and enquired boyishly after some sponge cake. Muck has been in the MacEwen family for 108 years, bought by Lawrence's great uncle who had earlier bought Eigg with money he'd made selling Armstrong battleships to the Japanese. As a child Lawrence and his brother Ewen were sent away to Gordonstoun where they were teased unmercifully; as Lawrence admitted, the address didn't help. Today the MacEwen family look after the island with a dedication and stewardship probably unmatched in the history of lairding. The family have made considerable efforts to plant trees and to rear lambs from 600 ewes and to raise calves from 50 cows. They grow their own vegetables, some in a polytunnel which, like most in the Hebrides after the January storm, was in shreds. The MacEwens are the kindest of people, extending hospitality to all visitors. Muck, like Canna, is an island to which I dream of returning.

canna, may 05 I had come to Canna to meet Gerry MacKinnon who is, by all accounts, an outstanding farmer. Gerry, with her sister Winnie and brother Patrick, all own their land on Sanday, a tidal island a short hop from Canna that would normally be accessible by bridge, but that was also wrecked in the January storm. The MacKinnons grew up on Canna with Gerry's father working as farm manager until he died in 1992. Gerry rears beef for Marks and Spencer and lambs from Cheviot sheep. It's a logistical struggle on all the Scottish islands to get livestock to market: they have to be herded onto military-style landing craft. We met on the hills where she kept her flock. I saw the rough land she'd set aside and heard the clear call of a corncrake hiding in some nettles. Meanwhile hooded crows glided under the steep cliffs and buzzards watched from the deepening storm clouds. Gerry and partner Murdo Jack had braved the foul weather to clean the bottoms of several hundred sheep. "I hate seeing sheep with dirty bums," remarked Gerry, wrapped in a wool hat and several layers of clothing. Ah, the romance of a Scottish island! At least the midges hadn't arrived.

faslane, june 05 Wishing to see our nuclear submarine fleet, I had contacted the Ministry of Defence ahead of my visit to the Clyde. I was told that the best way to see the submarines was by escorting them into harbour with the Clyde Marine Unit—police officers in launches. "We are the first line of defence," insisted Constable Mike Smith, cradling his Heckler and Koch MP5 submachine gun. The launch was patrolling outside the shed-like Trident nuclear missile store on Loch Long. The Ministry of Defence Police support Britain's defence against terrorist attack on naval traffic and installations around the coast. In the Clyde their main concern is small boats loaded with explosives moving into the path of naval ships and submarines, especially the Trident nuclear fleet. So far the worst they've had to deal with is anti-nuclear protestors swimming under the black rubber booms that protect the Tridents to paint slogans on the hull. The Clyde was a very different place from the outer islands. Here I saw the secret arrival of a Dutch submarine, back from anti-drug trafficking duty in the Caribbean. There were several American frigates with soldiers manning .50-calibre machine guns. So near and yet so different from the peaceful anchorage on Loch Fyne I'd enjoyed on *Irini* the previous evening.

FORMBY POINT, MERSEY ESTUARY, 2005

FORMBY POINT, MERSEY ESTUARY, 2005

FORMBY POINT, MERSEY ESTUARY, 2005

SEA-WATCHERS SPOTTING BIRDS AT FORMBY POINT, MERSEY ESTUARY, 2005

SANDWICH TERN EGGS, STRANGFORD LOUGH, NORTHERN IRELAND, 2005

NUGENT'S WOOD, STRANGFORD LOUGH, NORTHERN IRELAND, 2005

PORTAFERRY, STRANGFORD LOUGH, NORTHERN IRELAND, 2005

RIVER MERSEY, LIVERPOOL, 2005

CLYDE MARINE UNIT, RIVER CLYDE, SCOTLAND, 2005

DUTCH SUBMARINE ENTERING FASLANE, RIVER CLYDE, SCOTLAND, 2005

SCHOOL BUS, ARGYLL, SCOTLAND, 2005

WIND FARM, ARDROSSAN, SCOTLAND, 2005

DISUSED SLATE QUARRY, EASDALE, SCOTLAND, 2005

BLADDERWRACK, EASDALE, SCOTLAND, 2005

FERRY TO EASDALE, SCOTLAND, 2005

PUFFINS ON LUNGA, TRESHNISH ISLANDS, SCOTLAND, 2005

PUFFINS ON LUNGA, TRESHNISH ISLANDS, SCOTLAND, 2005

FINGAL'S CAVE, STAFFA, SCOTLAND, 2005

THE FALLS OF LORA ON LOCH ETIVE, ARGYLL, SCOTLAND, 2005

BASALT COLUMNS, ISLE OF STAFFA, SCOTLAND, 2005

LAWRENCE MacEWEN, LAIRD OF MUCK, ISLE OF MUCK, SCOTLAND, 2005

ISLE OF CANNA, SCOTLAND, 2005

GERRY MACKINNON, ISLE OF CANNA, SCOTLAND, 2005

ROSS OF MULL, SCOTLAND, 2005

EROSION AFTER 11TH JANUARY STORM, SOUTH UIST, SCOTLAND, 2005

EROSION AFTER 11TH JANUARY STORM, SOUTH UIST, SCOTLAND, 2005

technical note

The photographs in this book were taken on a Rolleiflex 3,5f (Whiteface) with 75mm f3.5 Planar, Hasselblad 500 CM and ELX with 60, 80 and 120mm lenses, Leica M6 and MP cameras with 28mm, 35mm and 50mm lenses. Film was mostly Fuji ACROS 100 ASA processed in Rodinal. Scans were made on an Imacon Flextight 848 scanner. The pictures of the Centenary Ball (including the cover image) were take on a Canon EOS 1Ds MarkII digital camera using a Canon 35mm f1.4 lens and strobe.

acknowledgements

I wish to acknowledge the enormous support I have received from the National Trust, and particularly from Maggie Gowan and Gaynor Aaltonen without whom, honestly put, this project would not exist. Along the way I have found kindness and support from National Trust staff, from the most senior advisors and managers to the volunteers manning the car parks. In particular, I would like to thank Rob Jarman, Phil Dyke and Nigel Hester. I'd like to thank Sarah Williams and Julie Fletcher for their skilled sailing help and good company and Mari Girling and Gaynor Aaltonen (again) for much needed help with text.

I'd like to thank the team at Magnum Photos, London—especially Sarah Rumens, Bree Seeley, Fran Sears, Brigitte Lardinois, Sophie Wright, David Axelbank, Paul Hayward, Alex Yallop, Sean Gallagher, Tanya Pfoehler and Ileana Athanatos—for their help on this project; the team at Bardwell Press—especially the designer Toby Matthews and publisher Peter Hamilton; Steve MacCleod at Potosi for the black and white printing; Nick Franklin for the scanning.

Also, I'd like to express my thanks to Rob Need, Andy Withers, Ben Sampson, Derek Green, Tim Vinnicombe, Danielle Rush, Neil Brockman, Lt Cdr Gareth Mawdsley, Lt Cdr Ahmed Ajala, Joy Prentice, Captain Russell Best, Ali Watkinson, Tony Barclay, Pete McHugh, Winnie and Gerry MacKinnon, the MacEwen family, Pete Smithies, Neil Smith, Richard Edmonds, Dr Margaret Storrie, Murdo Mackenzie, Willie Ramsay, Matthew Chatfield, Neil Peters, Emma Wright, Barry Ashdown and Dawn Savage and Paul and Lisa Dartnall.

Finally, I'd like to thank my family for their support during long absences.

Stuart Franklin
August 2005